Once THERE WERE NO COWS

Original oil paintings and story by

Bonnie Mohr

TO
MY HUSBAND, JOHN,
YOU ALWAYS BELIEVED IN ME, MORE THAN I DID IN MYSELF!
YOU ARE THE GREATEST DAIRY FARMER IN THE WORLD.

FOR
MOM AND DAD
THANKS FOR RAISING ME RIGHT...ON A DAIRY FARM,
AND FOR TRIXIE, MY FIRST COW.

Thank you to the Bonnie Mohr Studio and Hoard's Dairyman teams for their support and assistance in preparing this book for publishing.

The images in this book were created by artist Bonnie Mohr using oil paints.
Many of these original oil paintings are available as prints. Please visit us online to learn more about artist Bonnie Mohr or to purchase her originals and reproductions of Rural American & Inspirational artwork at

WWW.BONNIEMOHR.COM

"THE COW IS THE FOSTER MOTHER OF THE HUMAN RACE – FROM THE DAY OF THE ANCIENT HINDOO TO THIS TIME HAVE THE THOUGHTS OF MEN TURNED TO THIS KINDLY AND BENEFICENT CREATURE AS ONE OF THE CHIEF SUSTAINING FORCES OF HUMAN LIFE."

W.D. Hoard

Co-published by:
Bonnie Mohr - www.bonniemohr.com
W.D. Hoard & Sons Company - www.hoards.com

Printed in the United States of America

Library of Congress Control Number: 2014944528
Mohr, Bonnie L.
Once There Were No Cows / Bonnie L Mohr
ISBN 978-0-9960753-1-2
18 17 16 15 14   1 2 3 4 5
First Edition

ONCE THERE WERE NO COWS

Dedication

To all dairy farmers -
who devote many selfless hours of love, dedication
and caring for their cows,
every single day of the year.
All while enduring working conditions of
harsh weather, unpredictable prices, and long, long hours -
but they keep on doing it anyway,
because they love what they do.
To the producers of nature's most nearly perfect food
that gets picked up by the milkman,
and hauled to the creamery
to bc bottled and processed,
to help feed the world,
and is then trucked to the grocery store,
to be purchased by shoppers,
and brought home,
to be poured onto Cheerios,
dunked by Oreos,
or whipped into a smoothie.

Thank you
for the many people you help to employ,
for working outside when it is 20 below,
for all the new baby calves you deliver, in the middle of the night,
for raising such great kids,
for helping to promote June Dairy Month,
and most of all,
for living life as a guardian
to God's most wonderful creature, the cow.

Thank You.
This book is especially for you.

Once there were no cows . . . . .

There was no life at all. There was nothing - but God.

The universe had no form. It was an empty, cold, dark place. God looked around and saw nothing. He was full of love but felt a little lonely. He wished to share His love and began to wonder . . . what if? So God made a plan, and time began!

Amid the dark, God spoke, and the sky lit up with the sun, the moon and stars! Now He could eyeball the situation, and this was much easier than working in the dark. (This is where the phrase "it brought a whole new light to the situation" came from!) God was ready to create life, but first, He needed a place to put everything, a place to call "home." And so He made the world.

The earth was a grand beginning - but perhaps the solid brown sphere was rather dreary - so God imagined color. "This will be a marvelous touch," He thought. So He painted a brilliant blue sky, and added sparkling, crystal clear water below, to reflect its cerulean color.

Then He added ground and called it land. God smiled and continued on wit
His plan, producing plants, trees, fruits, and vegetation for eating and feedin

God saw that all these things were good and that He had made a splendid atmosphere and inviting landscape.

God longed for companionship
and life that would dance and play
and roam on the beautiful world He
created, and so He blessed the earth
with animals and creatures that swam
and flew and traveled about.

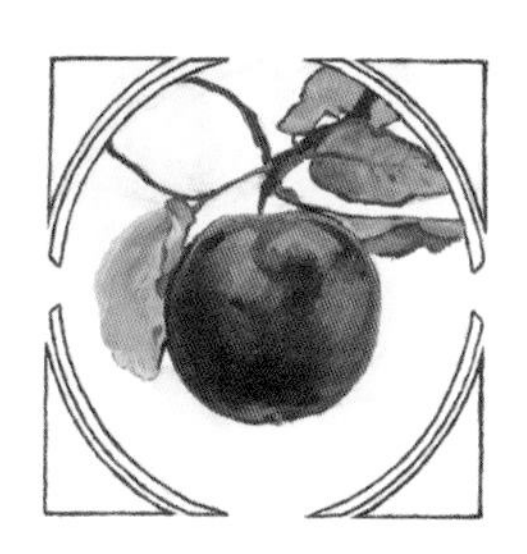

God desired human companionship, too - someone to care for His creatures and be a guardian to watch over the earth. So God breathed deeply and using dust from the ground . . . He formed Adam and Eve. What a happy day this was!

God's big project was wrapping up and He saw that it was all wonderful, but He noticed that among all the new creatures, there was something missing. There was no animal that demonstrated the abilities or provided the qualities that man needed to become a thriving steward of the earth.

God wanted an animal that would give more than it would take. An animal that would work as hard as man, offering him a foundation on which to raise a family and teach his children about responsibility and reward, to give them strong roots and prepare them for life.

God wanted an animal that would be appreciated and loved intensely because of the profound impact she would have on all mankind . . . a living thing that would bring out the best in man and make him smile. One that would provide a livelihood of toil and labor that could be both enjoyable and meaningful.

And so, in God's final moments of creation - He formed the cow. In her majesty and loveliness, she would be equally productive, providing both meat and milk to feed the world. Of all animals, she would be a sustainable force on earth, providing much to all mankind. And so it was, one by one they appeared. (God decided that it would be best to arrive in alphabetical order.) First came the Ayrshire . . .

. . . . . and then the Brown Swiss.

Next followed the Guernsey.

Shortly after arrived the Holstein . . . . .

. . . . . and then the Jersey . . . . .

. . . . . and finally, the Milking Shorthorn!

God sat back and smiled. He was joyful and pleased. He knew the world was now complete with His beautiful cow. Because her harmony with God was deemed of the highest level she would be named the Foster Mother of the Human Race.

Dairy farming would be a special and beautiful lifestyle chosen people would embrace. It would require hard work and long days but would provide a good life, filled with rich and meaningful lessons. Dairy farmers would enjoy families and children who would be God's most tender and joyous people and have true appreciation for the simple pleasures in life. Above all, they would be blessed with great love . . . in the house, in the barn and threaded deeply into their hearts.

Six days had passed and God's magnificent creation of the world was now finished. And so, on the 7th day He rested for everything was good and right. And then God said, "Now it is time to start *Living Life!*"

The Beginning . . .

Bonnie Mohr has a love and passion for cows, and she has enjoyed a career of painting them for over 25 years. One of the world's most renown bovine artists now authors her first book to accompany her stunning, life-like paintings in this story of creation and the cow. A self-taught artist, Bonnie grew up on a dairy farm near St. George, Minnesota. Bonnie and her husband, John, currently reside on their own dairy farm near Glencoe, Minnesota, where they have raised five children. Here she operates Bonnie Mohr Studio, which is open to the public and tour groups. Her Rural American and Inspirational artwork continues to capture the essence and beauty of the simple moments in life, while serving as a voice and advocate of the dairy industry and rural living.

WWW.BONNIEMOHR.COM